Essential Question

What are the structures of plants and their functions?

Plants

by Daphne Greaves

Pixtal/age fotostock

Chapter 1

A Feast for the Senses

You are in the United States Botanic Garden on the National Mall in Washington, D.C. Breathe deeply and smell the sweet scent of flowers and the sharp tang of tree sap. Reach out and brush your fingers across velvety moss. Look around at vivid colors of yellow, red, and orange. Enjoy the cool shades of purple, lavender, and blue. Relax to the calming shades of green. Exhibits of vegetables, fruit trees, and spices might even make you hungry!

A wide variety of plants can be found at the United States Botanic Garden in Washington, D.C.

The 26,000 plants in the United States Botanic Garden are a feast for the senses and a reminder of the importance of plants in your life. In addition to being beautiful, plants are a source of food and medicine. Their wood is used to make furniture and build houses. Their fibers are used to make paper and cloth. Most importantly, plants play a vital role in a process that produces the very air we breathe. Without plants, most life on Earth could not exist.

Visitors to the United States Botanic Garden Conservatory explore the beauty of its many plant species.

(t) Lane Oatey/Blue Jean Images/Getty Images; (b) PNC/Stockbyte/Getty Images

Chapter 2

Plants and Parts

Most plants are rooted in the ground. They produce their own food and grow practically everywhere. Plants are grouped into two main categories, flowering and nonflowering. Roses, geraniums, and maple trees are just a few of the many flowering plants. Nonflowering plants include mosses, ferns, and pine trees. Most plants have two primary types of parts, a root system and a shoot system. The root system is made up of the plant's underground roots and root hairs. The shoot system includes the plant's stems, leaves, and reproductive system.

Ferns are a kind of nonflowering plant.

Gerber daisies are flowering plants.

Most plants have one of two types of roots: taproots and fibrous roots. A taproot has one large main root with smaller branch roots growing from it. A carrot is an example of a taproot. Plants with fibrous roots, such as grasses, have many small roots that are all about the same size. These roots do not grow as deeply into the soil as taproots. Some plants, like many trees, start life with a taproot but change to a fibrous root system as they grow.

Carrots have taproots.

Grass has fibrous roots.

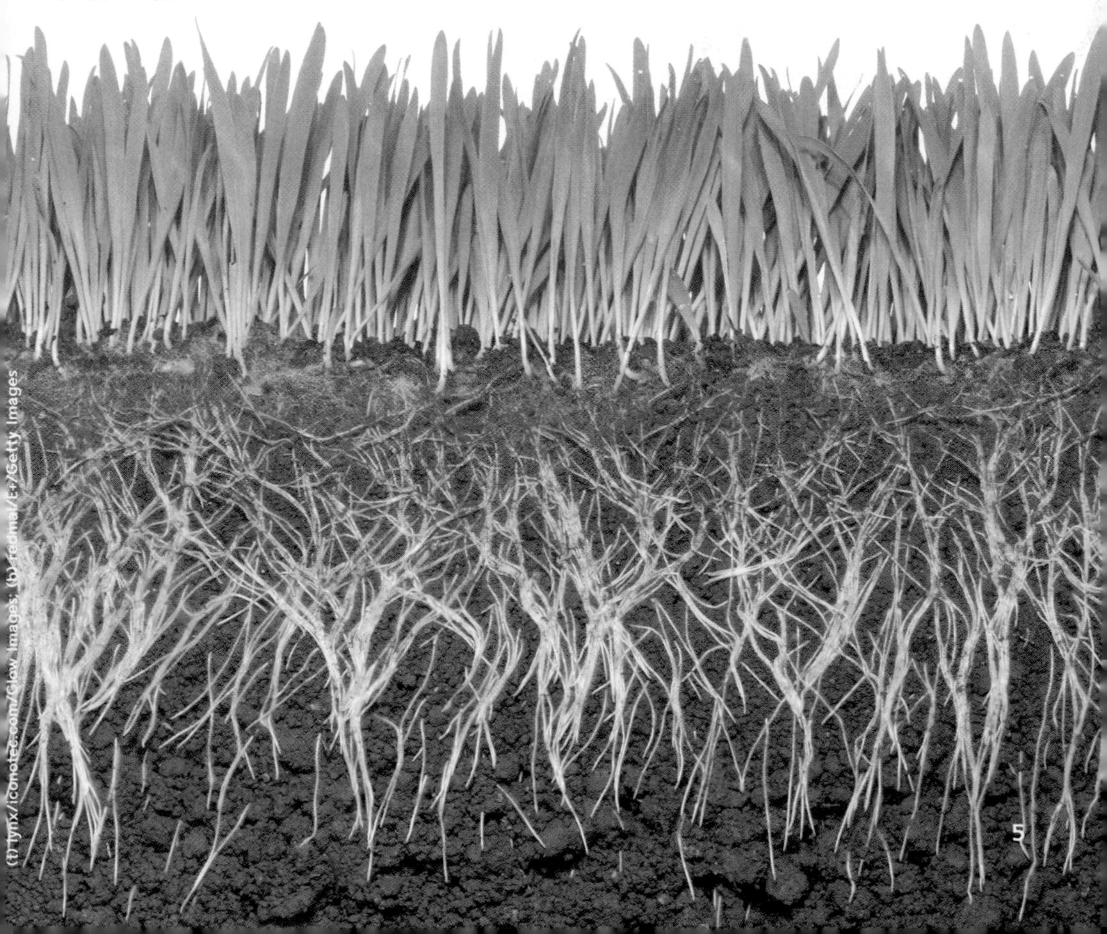

Both taproots and fibrous roots grow underground. Their ends are rounded and covered by a root cap. This cap covers the root to protect it and keeps the root from being cut and scratched as it pushes through the soil.

This magnified corn root is cut to show the root cap.

In addition to holding the plant in place, the roots absorb water and nutrients from the soil. Roots are covered in tiny root hairs. These root hairs increase the surface area of the root, providing more contact with the soil. This allows roots to take in greater amounts of water and nutrients. After being absorbed, nutrients and water are stored in the roots and are kept there until they are needed by other parts of the plant.

The root hairs of this carrot absorb water and other nutrients from the soil.

A plant's stem rises up above the ground from the roots. For trees, the trunk and branches are the stem of the plant. Most plant stems grow vertically, or upward, toward the Sun. Some plant stems grow horizontally, such as vines. Strawberries, for example, have stems called runners that spread out horizontally from the main part of the plant. Stems support the plant's leaves, whether they grow vertically or horizontally.

Leaves are attached to the plant stem by a **petiole**. The petiole is a smaller stem that connects the plant's main stem to a leaf. The outer surface of a leaf is protected by a waxy coating called a cuticle. Inside each leaf, veins carry water and nutrients. Leaves get their green color from a pigment called chlorophyll.

Leaf Shapes

Plant leaves come in many different shapes and sizes. A simple leaf is made of a single leaf blade and a petiole. Simple leaves can be long and thin or round and fat. Leaves can be oblong (longer than they are wide), triangular, or heart shaped. Sometimes a single plant will have leaves with different shapes.

Compound leaves are made of many smaller leaves, called leaflets, attached to a single petiole. Leaflets can attach to the petiole at one point, like a fan, or they can attach along the length of the petiole and look like many leaves on a branch.

Leaf Shapes

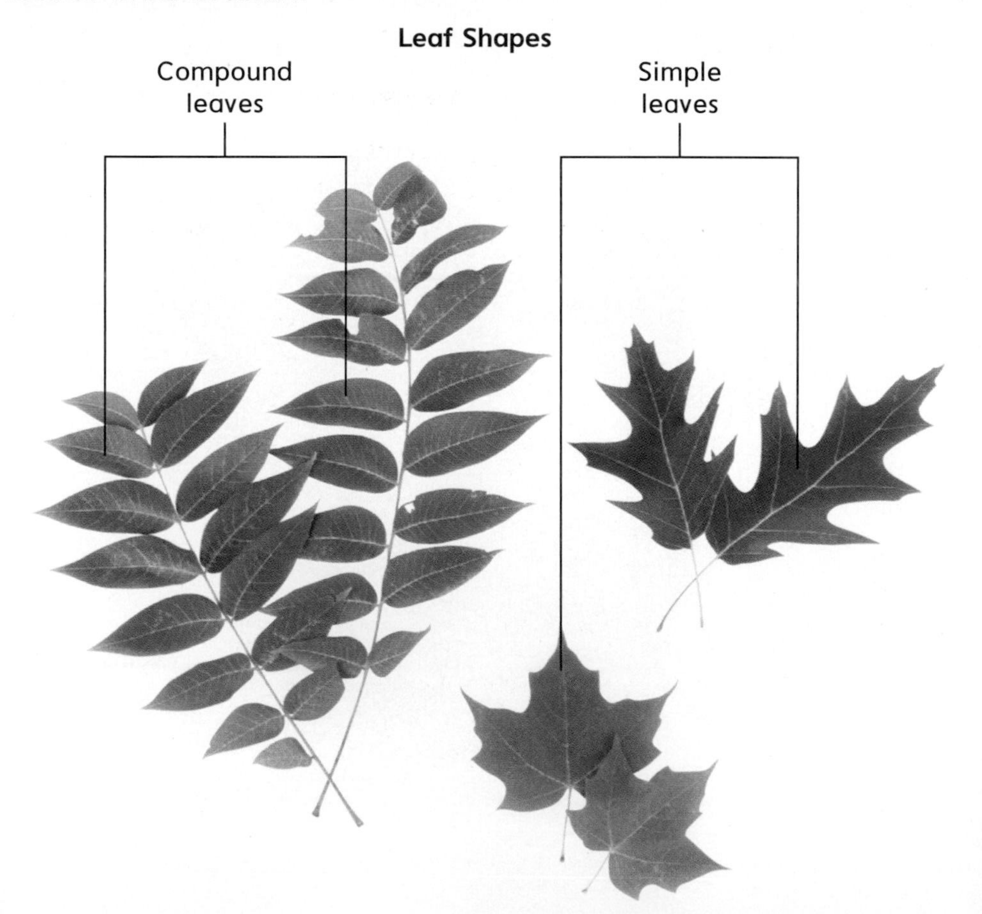

Flowers come in many different shapes and colors, but all flowers share the same basic parts. In the center of the flower are the plant's male and female reproductive parts. The female part is called the pistil and eventually becomes a seed. The male part is called the stamen and produces pollen. In a process called pollination, pollen from the stamen enters the pistil. Successful pollination makes fertile seeds from which new plants grow.

The male and female parts are protected by the flower's petals. The colors, shapes, sizes, and scents of petals vary greatly and attract bees and other insects. When an insect lands on a flower, pollen from the stamen rubs off onto the insect's body. The insect then moves to another plant, leaving behind pollen as it goes. The pollen fertilizes the pistil, enabling new flowers to grow. In this way, flowers encourage insects to play a helpful role in pollination.

This illustration shows the male and female parts of a flower.

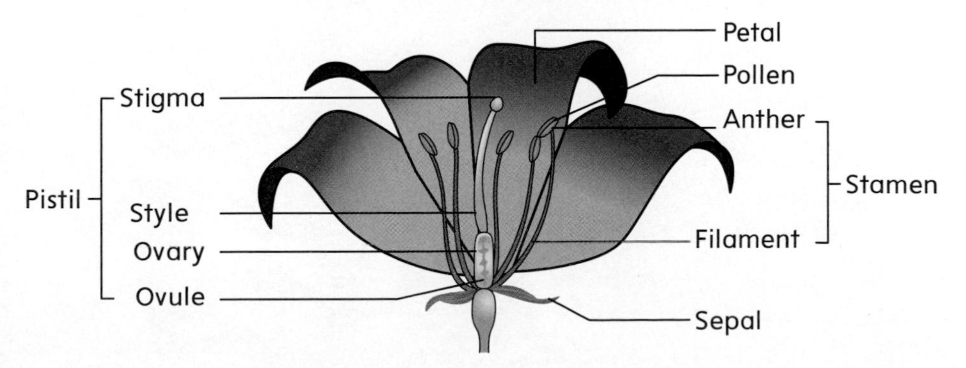

Chapter 3

Photosynthesis

Photosynthesis is an important process for all life on Earth. Plants use photosynthesis to make their own food and to provide food for animals. Animals that eat plants get their energy directly from plants. Other animals get energy from plants indirectly by eating the animals that eat plants. As shown in this energy pyramid, life on Earth is nourished by the photosynthesis of plants.

Life depends on the Sun's energy.

(bkgd) sanjagrujic/iStock/Getty Images; (t-b) mallardg500/Getty Images; Mark Dierker/McGraw-Hill Education; Thanks for visiting/Moment Open/Getty Images; STOCK4B-RF/Glow Images; (bl) Jezperklauzen/iStock/Getty Images; (br) Paulina Lenting-Smulders/Getty Images

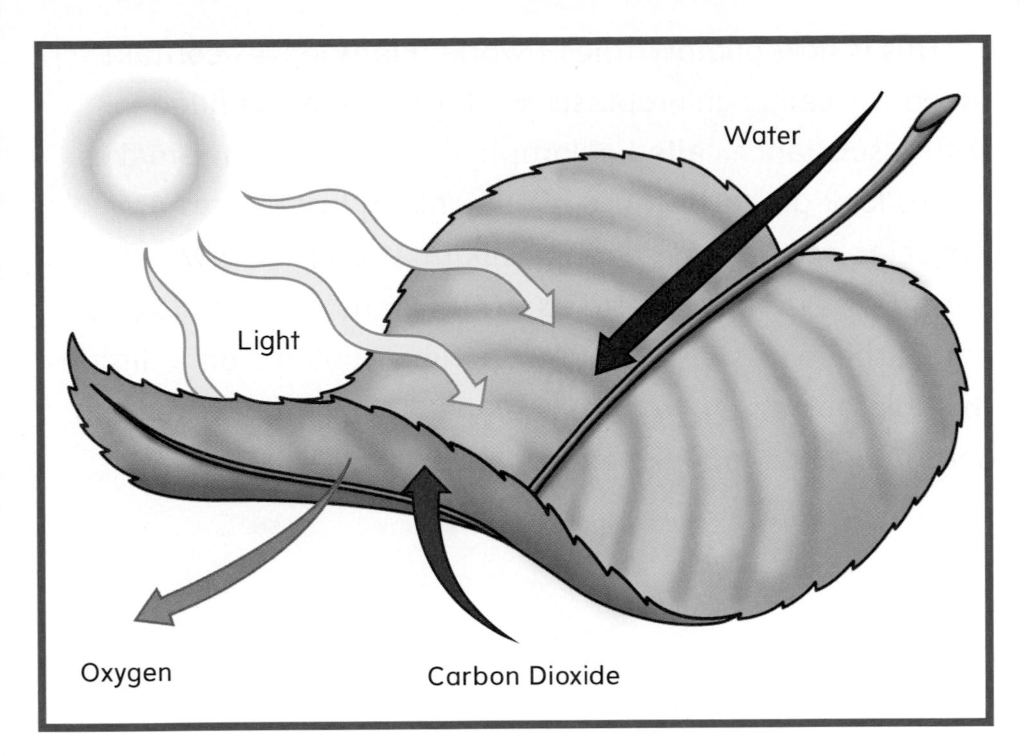

Photosynthesis uses the Sun's energy to make oxygen and glucose from water and carbon dioxide.

Plants also help make the air we breathe by producing oxygen. Oxygen makes up 21 percent of the air. It is produced and released into the atmosphere during photosynthesis. Without it, life on Earth would not be possible.

The process of photosynthesis uses sunlight to change water and carbon dioxide into glucose and oxygen in the leaves of plants. Glucose, a form of sugar, is a source of energy for plants and animals, giving them energy to perform necessary functions.

This is how photosynthesis works. Plant leaves contain tiny parts called chloroplasts. Each chloroplast is filled with a substance called chlorophyll, a pigment that makes leaves look green. Because of the chlorophyll, leaves are like solar panels that collect sunlight. Water is drawn up into the leaves from the roots of the plant. Carbon dioxide enters the leaves through tiny openings called stoma. Inside the chloroplasts, energy from sunlight acts upon water and carbon dioxide to make oxygen and glucose.

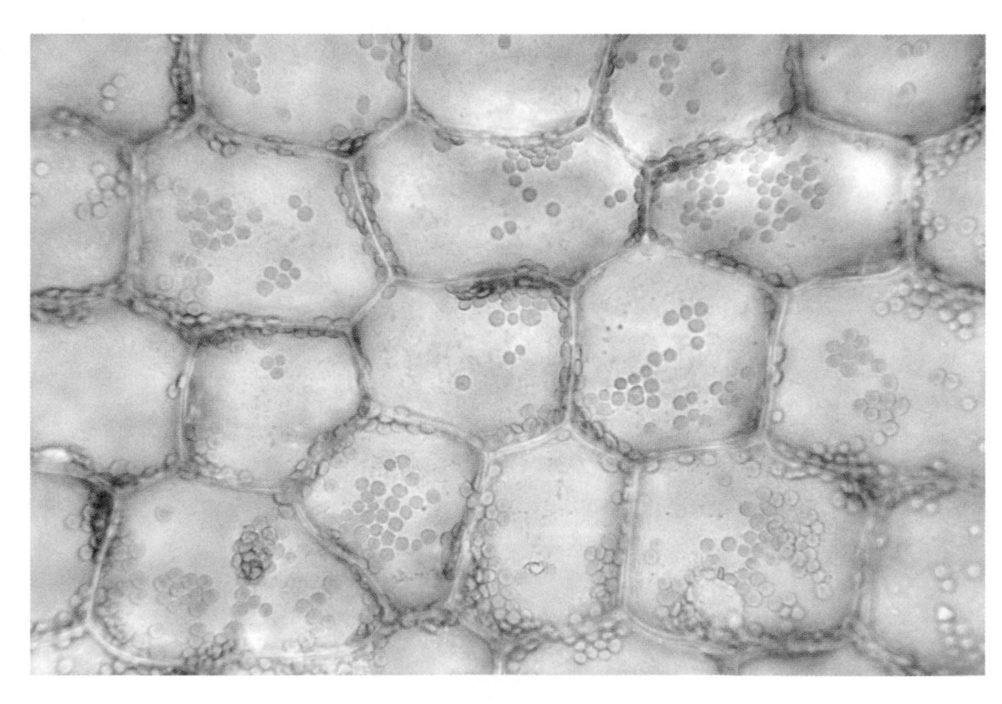

Nancy Nehring/E+/Getty Images

The green spots in these plant cells are chloroplasts. Each chloroplast is filled with chlorophyll.

Chapter 4

Cellular Respiration

The glucose that plants make during photosynthesis is a source of energy. For plants and animals to benefit from this energy, they must change the glucose into a substance called **ATP**. ATP is an energy source that is then used by all living things to grow and function. **Cellular respiration** is the process that breaks down glucose into ATP.

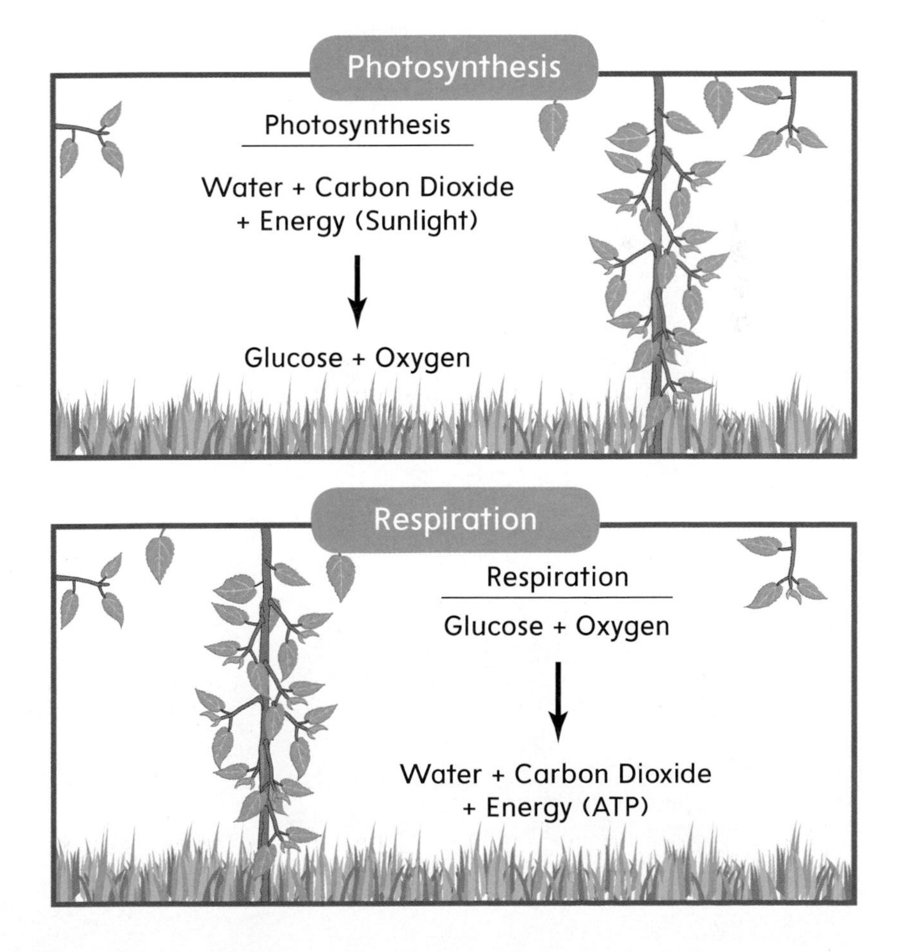

In cellular respiration, glucose molecules are broken down into atoms inside plant cells. The atoms from the glucose molecules are then combined with oxygen atoms. Combining glucose with oxygen releases the energy source ATP. **By-products** of this process are carbon dioxide and water.

The energy from cellular respiration will help this seedling grow into a tree.

(t) ©Steven P. Lynch; (b) Exactostock/SuperStock

Chapter 5

Moving Water and Nutrients in Plants

Plants absorb water and nutrients through their roots and move it to their leaves to be used in photosynthesis. They also need to move the glucose produced in the leaves to all the living cells in the plant. Water, nutrients, and glucose are transported throughout plants by a system of tubes that form the **vascular system**. These tubes are made up of special cells called xylem cells and phloem cells.

Cross-section of a young lime tree stem sapling under a light microscope.

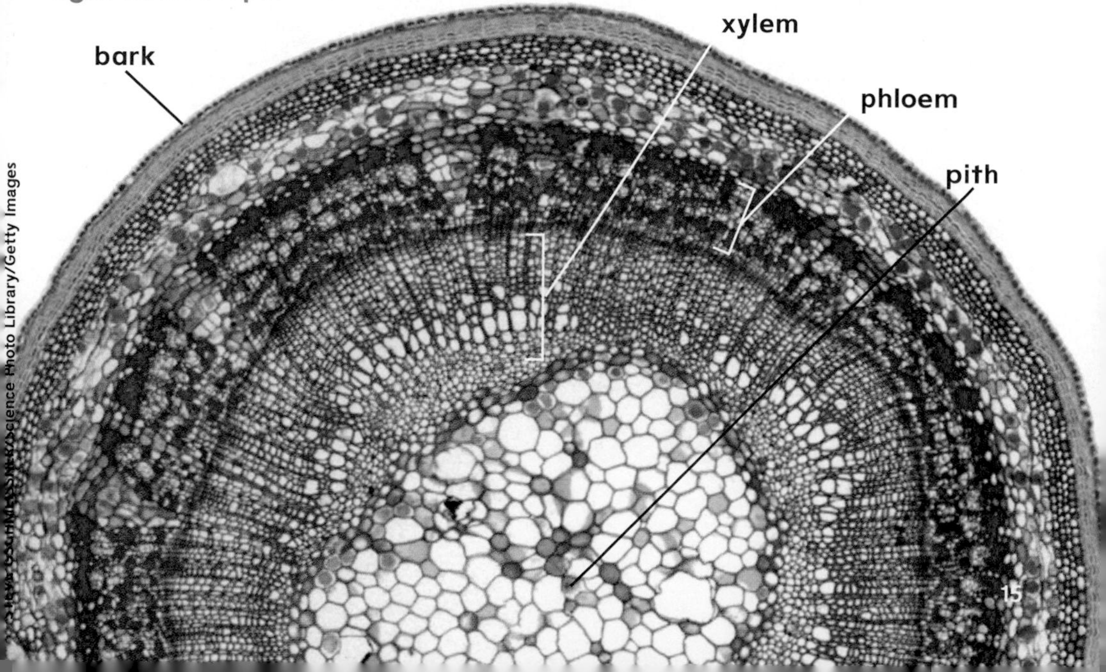

Xylem cells line up throughout the plant, forming long chains of cells. When they are fully grown, the nucleus and other parts of the cell die, leaving a hollowed-out tube. Fluids are then able to move through the tube. Xylem cells carry water up from plant roots to stems and leaves. Xylem cells are also rigid and help support the plant. The wood of trees is made of xylem cells.

Phloem cells move glucose from the plant's leaves to its other parts. They also distribute ATP produced during cellular respiration. In trees, the phloem is part of the bark.

Cross-section of a nasturtium under an electron microscope

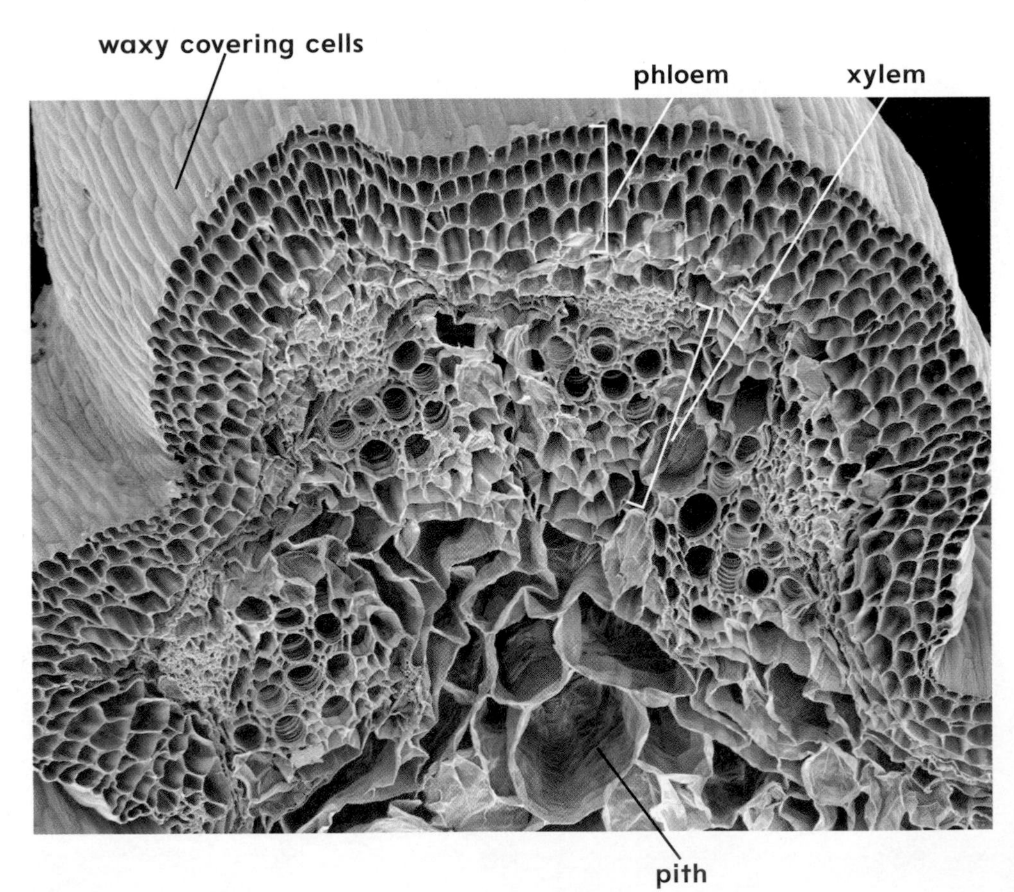

STEVE GSCHMEISSNER/SPL/Getty Images

Chapter 6

How Plants Remove Wastes

All living things must remove wastes to stay healthy. The main way that plants remove wastes is by storing them in a **vacuole**. A vacuole is a balloon-like container found in all plant cells that acts like a trash can. There, harmful wastes are kept away from the rest of the plant. Plants can also store wastes in their leaves. When leaves die and fall away, the wastes stored there are removed from the plant.

This drawing shows a large central vacuole in a plant cell.

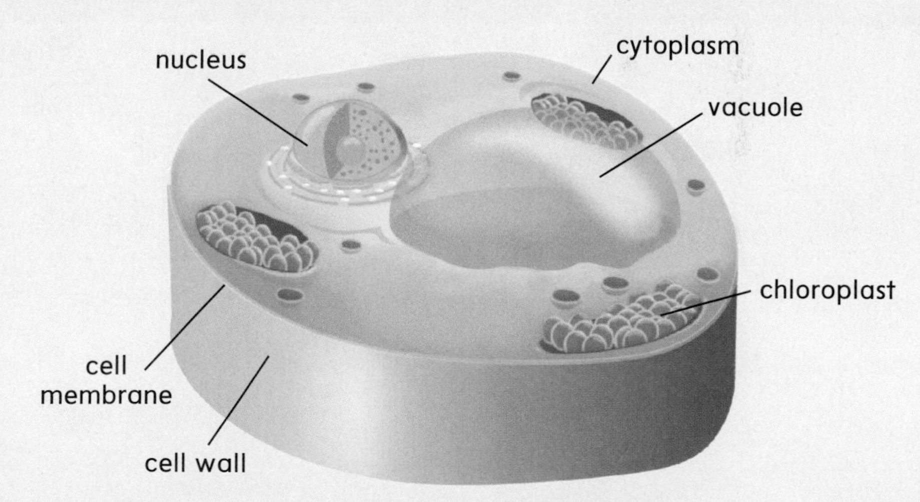

Storing wastes in vacuoles instead of getting rid of them can make a plant taste bitter. This can provide protection for the plant by discouraging insects from eating it. Some plants release wastes into the soil, making the soil harmful to other kinds of plants.

Each moss cell has a large vacuole and many chloroplasts.

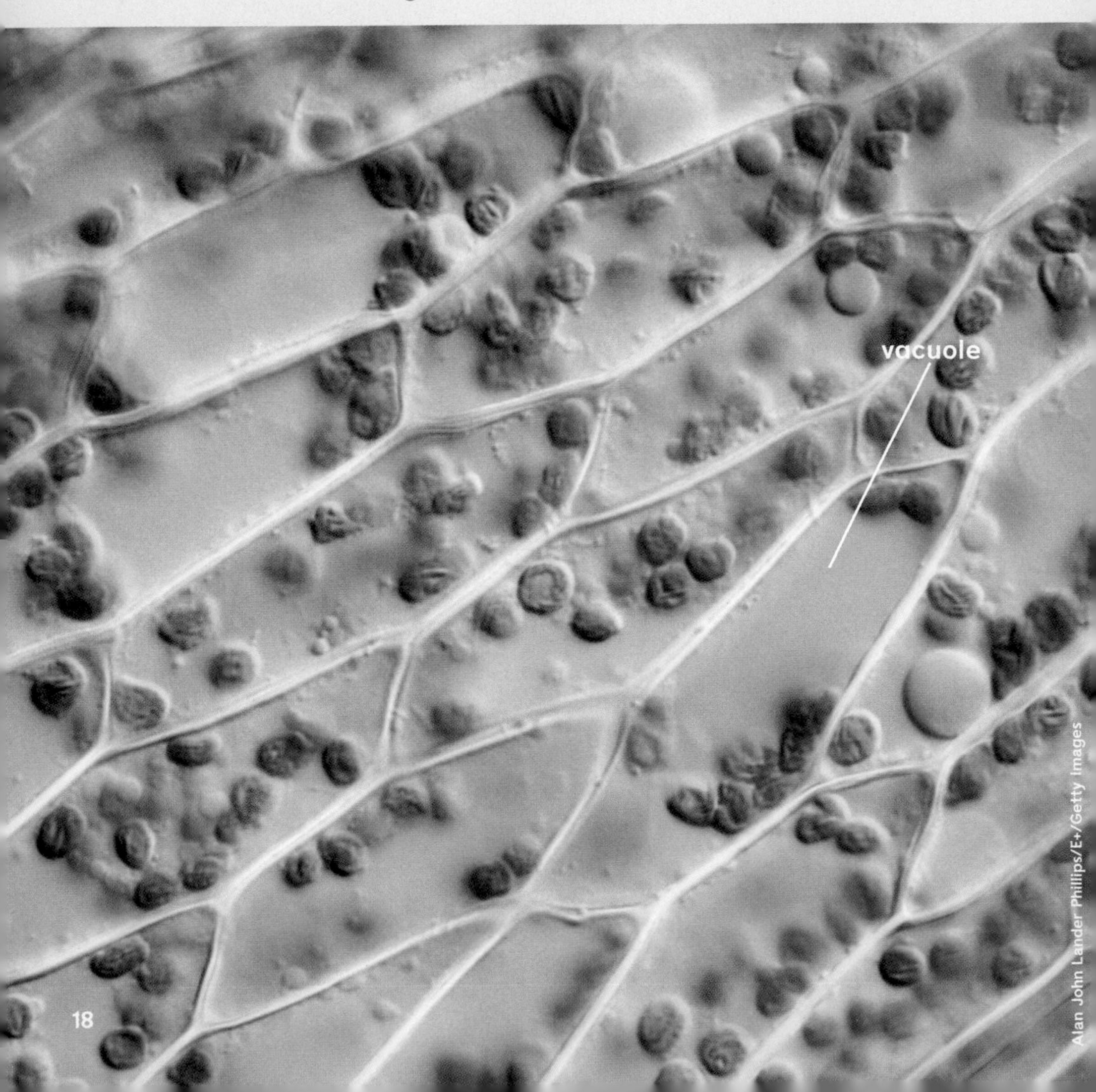

Chapter 7

Plants and People

In early human history, people gathered wild plants for food. Later, people domesticated plants for human use. Through planned breeding, people have made plants easier to grow and improved the taste and nutritional value of crops. People also breed plants to grow prettier flowers that live longer. Today, we can trace back some domesticated plants thousands of years to their wild ancestors.

Every day, botanists study and learn more about plants. Through their work, human knowledge of plant structures and functions continues to grow. Botanists teach everyone how important plants are to all life on Earth.

An ancient Egyptian man is shown plowing a field. The Egyptians knew how to grow plants.

A botanist examines a plant.

(l) udra/iStock/Getty Images; (r) Javier Larrea/Pixtal/age fotostock

Summarize

Use important details from *Plants* to summarize the selection. Your graphic organizer may help you.

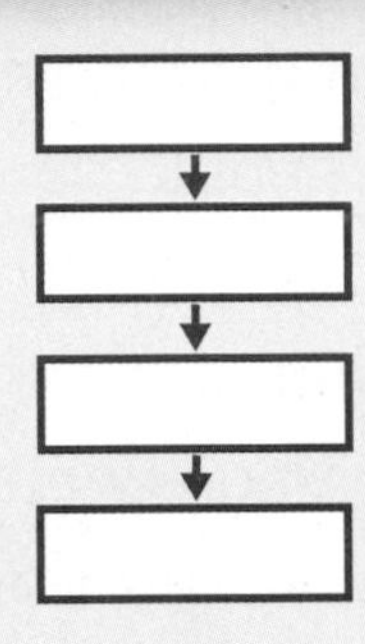

Text Evidence

1. What are main functions of a plant's root system?

2. Read the book again with a small group. Write a summary of photosynthesis. List the steps of the process in order. SEQUENCE

3. What does *hollowed-out* mean on page 16? Which context clues helped you figure out the meaning? CONTEXT CLUES

4. Write a summary of cellular respiration. Begin with a topic sentence and list the steps of the process in order. Include only important details, such as what the process is, how it works, and what plant structures are involved. Share your report with your class. WRITE ABOUT READING

Genre Fiction

Compare Texts

Read about how Aisha helps her parents grow plants on Mars, a world without plants.

A World Without Plants

"Come on, Aisha," called her mother. "It's time to get into your spacesuit. This is the big day!"

Aisha looked out of the clear dome that formed her Martian home. The sky, as usual, was red and cloudless. "Be right there," she called back.

Aisha thought back on the incredible events that had brought her here. Who would have thought that having botanist parents would earn you a trip to Mars? Of course, plants were the key to colonizing the red planet.

"Mars is a world without plants," her mother had told her. "No plants means no air, no food, no clouds, no rain . . . We've already gotten plants to grow in Martian soil. The question we have to answer now is, "Can plants grow in Martian gravity, and so far from the Sun?"

When they'd arrived on Mars, Aisha and her parents had worked to set up a dome and plant different types of crops on the Martian plains. They'd worked hard to give the plants a good start. Today, they'd take measurements to find out whether their work would bear fruit.

Aisha's father checked her helmet to be sure it was safely secured. She turned on her air supply. The airlock opened, and the family climbed aboard their rover. In just a few minutes, they'd arrived at the edge of their test beds.

While Aisha gathered samples of each different type of plant—soy, corn, tomatoes, young trees—her parents made measurements. The plants were surviving, but would the measurements show that the experiment would succeed?

Back at the dome, Aisha waited impatiently for her parents to finish their measurements. The lab door opened, and her mother stepped out. "Congratulations," she told Aisha. "We did it! We're here to stay. It's up to us to lead the way to colonizing a green, livable Mars!"

Make Connections

Why was it so important for Aisha and her parents to successfully grow plants on Mars?

TEXT TO TEXT

Glossary

ATP *(AY-tee-pee)* an energy source for plants made when atoms from glucose molecules are mixed with oxygen atoms *(page 13)*

by-product *(BIGH-prod-ukt)* something useful that results from the making of something else *(page 14)*

cellular respiration *(SEL-yuh-luhr res-puh-RAY-shuhn)* the process that breaks down glucose into an energy source used by plants *(page 13)*

petiole *(PET-ee-ohl)* a smaller stem that connects the plant's main stem to the leaf *(page 7)*

photosynthesis *(foh-tuh-SIN-thuh-sis)* a chemical process by which green plants trap light energy to change carbon dioxide and water into carbohydrates (sugars) and oxygen *(page 10)*

vacuole *(VAK-yew-ohl)* the cell's holding bin for food, water, and waste *(page 17)*

vascular system *(VAS-kyuh-luhr SIS-tuhm)* a system of tubes that transports water and nutrients through plants *(page 15)*

Index